The Cell Manufacturing Playbook

A Step-by-Step Guideline for the Lean Practitioner

The LEAN Playbook Series

The 5S Playbook: A Step-by-Step Guideline for the Lean Practitioner
Chris A. Ortiz

The Kanban Playbook: A Step-by-Step Guideline for the Lean Practitioner
Chris A. Ortiz

The Quick Changeover Playbook: A Step-by-Step Guideline for the Lean Practitioner
Chris A. Ortiz

The TPM Playbook: A Step-by-Step Guideline for the Lean Practitioner
Chris A. Ortiz

The Cell Manufacturing Playbook: A Step-by-Step Guideline for the Lean Practitioner
Chris A. Ortiz

The LEAN Playbook Series

The Cell Manufacturing Playbook

A Step-by-Step Guideline for the Lean Practitioner

Chris A. Ortiz

CRC Press
Taylor & Francis Group
Boca Raton London New York

CRC Press is an imprint of the
Taylor & Francis Group, an **informa** business

A PRODUCTIVITY PRESS BOOK

CRC Press
Taylor & Francis Group
6000 Broken Sound Parkway NW, Suite 300
Boca Raton, FL 33487-2742

First issued in paperback 2019

© 2016 by Taylor & Francis Group, LLC
CRC Press is an imprint of Taylor & Francis Group, an Informa business

No claim to original U.S. Government works

ISBN-13: 978-1-138-43793-7 (hbk)
ISBN-13: 978-1-498-74170-5 (pbk)

Library of Congress Cataloging-in-Publication Data

Ortiz, Chris A., author.
 The cell manufacturing playbook : a step-by-step guideline for the lean practitioner / Chris A. Ortiz.
 pages cm
 Includes index.
 ISBN 978-1-4987-4170-5
 1. Production management. 2. Manufacturing processes. 3. Lean manufacturing. I. Title.

TS155.O768 2016
670--dc23   2015026074

Visit the Taylor & Francis Web site at
http://www.taylorandfrancis.com

and the CRC Press Web site at
http://www.crcpress.com

Contents

How to Use This Playbook

In most cases, a playbook is a spiral bound notebook that outlines a strategy for a sport or a game. Whether for a football game, a video game, or even a board game, playbooks are all around us and when written properly provide immediate and easily understood direction. Playbooks can also provide general information; then it is up to the user of the playbook to tailor it to their individual needs.

Playbooks contain pictures, diagrams, quick references, definitions, and often step-by-step illustrations to explain certain parts. You can use playbooks to help you understand the entire game or you can pick and choose to focus on one element. The bottom line is that any playbook should be easy to read and to the point and contain little to no filler information.

The *Cell Manufacturing Playbook* is written for the Lean practitioner and facilitator. Like a football coach, a facilitator can use this playbook for quick reference and be able to convey what is needed easily. If for some reason the person leading the actual work cell implementation forgets a "play," the playbook can be referenced.

You can follow page by page and use the playbook to facilitate a work cell implementation or you can go directly to certain topics and use it to help you implement that particular "play." As a side note, I use the terms *cell manufacturing* and *work cell* as the same in definition.

Introduction

At first glance, the improvement techniques within the Lean philosophy appear to provide a solution to many types of production-related issues. A powerful and effective improvement philosophy, Lean can prevent company failure or launch an organization into world-class operational excellence.

I have been a Lean practitioner for over 15 years and have been involved in many Lean transformations. It does not matter the industry you work in, the product you produce, or even the processes your company uses to transform something to a finished good, the problems and opportunities you face are the same as those of everyone else. Your company is not "different" or an exception. You, as a Lean practitioner, desire a smoother-running facility, reduced lead times, more capacity, improved productivity, flexible processes, usable floor space, reduced inventory, and so on. Organizations can implement Lean to make localized improvements or they can use Lean to transform the entire culture of the business. Regardless of your aspirations and goals for Lean, you and many other companies face another similar situation; getting out of what I call *boardroom Lean,* and moving toward implementation.

Have no illusions, Lean is about rolling your sleeves up, getting dirty and making change. True change comes on the production floor, in the maintenance shop, and in all the other areas of the organization and by implementing the concepts of Lean. Companies often become stuck in endless cycles of training and planning, with no implementation ever happening. This playbook is your guideline for implementation and is written for the pure Lean practitioner looking for a training tool and a guideline that can be used in the work area while improvements are conducted. There is no book, manual, or reference guide that provides color images and detailed step-by-step guidelines on how to properly implement a work cell. The implementation of cell manufacturing is a manually intensive action, and conducting work cell projects properly takes experience and direction. The *Cell Manufacturing Playbook* is not a traditional book, as you can probably see. It is not intended to be read like another Lean business book. The images in this playbook are from real work cell implementations, and I use a combination of short paragraphs and bulleted descriptions to walk you through how to effectively implement cell manufacturing.

Little or no time is wasted on high-level theory, although an introductory portion is dedicated to the 8 Wastes and Lean metrics. An understanding of wastes and metrics is needed to fully benefit from this playbook. I am not implying that high-level theory or business strategies lack value; they are highly valuable. This playbook is for implementation, so it will not contain filler information.

The Introduction covers the 8 Wastes of Lean and a description of what cell manufacturing is and its application to improving operational processes. The Introduction also covers the six Lean metrics that are recommended in each implementation. It sets the reader up for why cell manufacturing is a great Lean tool. The Introduction is not intended to provide an understanding of Lean but is more specific to cell manufacturing as a tool within Lean.

Chapter 1 will cover the importance of data collection by describing how to conduct time and motion studies. This baseline information is extremely valuable in establishing the proper cycle times for each assembly step and to identify waste reduction opportunities. Work cell balancing will be covered in Chapter 2. To ensure a smooth-running work cell, work content must be distributed as evenly as possible between workers. Chapter 2 will also walk you through how to calculate available working time and Takt time as well. Work cells do not come together properly unless you spend time in design. Each work cell is customized based on the cycle time, products, and volume. Workstations should be designed based on the specifics of the work content being performed. Chapter 3 will cover a variety of key implementation tools that should be decided prior to setting up the cell.

Chapter 4 walks the reader through a 4-day work cell implementation outlining the activities and detailed work needed each day to have a fully operational work cell in 4 days. Using real-life examples, pictures, and diagrams, the reader will see the work cell be created as each day is described. 5S and Kanban are discussed as well using those playbooks from the series as reference.

Chapter 5 covers the proper methods for creating work instructions, a vital component for supervising and operating a Lean work cell. The chapter provides examples of instruction with both images/pictures and three-dimensional drawings to illustrate the actions for the workers. It also shows pictures and diagrams of how to properly install them right in the workstation for point-of-use application and ease of use.

This chapter also covers how to design and build production control boards (PCBs) for the work cell. PCBs are visual target boards that provide a guideline and goals for end-of-day volume. PCBs monitor output and provide feedback in short time frames displaying the work cell's progress and issues. Multiple examples, and even how-to pictorial guidelines, are included. From design, construction, and implementation, this chapter provides a nice finishing touch to this great implementation training manual.

Cell Manufacturing

Cell manufacturing is a concept that takes traditional process-based production environments and creates right-sized smaller work areas. The smaller work areas have all of the essential workstations and equipment needed to produce the product.

Problems with Processed-Based Production

- **Processes are separated by distance:** This adds time and uses too much floor space.
- **Batch processing of high work in progress (WIP):** Excessive inventory potentially hides errors and wait time.
- **Long manufacturing lead time:** This is the time it takes to get through the whole process.
- **Limited visibility of problems:** People and processes cannot be seen by each other.
- **Reduced sense of urgency:** Excessive WIP is a work buffer.
- **Build to a ship date:** This does not take into account what the next process really needs.
- **Limits growth:** When volume increases, waste increases along with it.

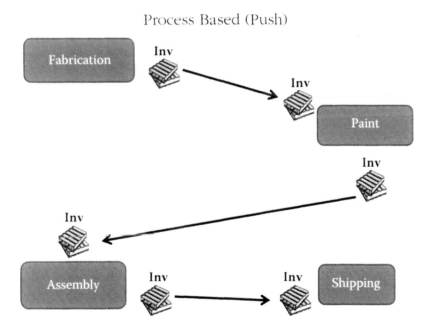

Process Based (Push)

Benefits of Cell Manufacturing

- Stations and machines are closer to each other: reduces distance and shortens lead time.
- Single-piece flow or small controlled batches: reduces lead times and requires less inventory.
- Cells are often U or C shaped: creates a work team environment.
- Quick changeovers: forces improvements to setup and changeover.
- Flexible environment for mix model: designed for higher mix.

- Reduced lead times and cycle times: creates less distance, time, and waiting.
- Greater visibility of mistakes: provides instant feedback and constant team communication.
- Elevated sense of urgency: eliminates time to build up unneeded inventory.
- No need for "silo"-style departments: creates right-sized workstations for each work cell.

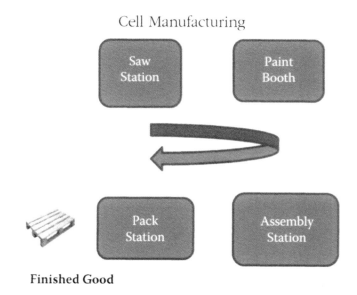

Cell Manufacturing

8 Wastes

As a Lean practitioner and teacher, I know the power of cell manufacturing; when you challenge your viewpoint and the handling of inventory, you can see remarkable changes to your overall company. Like any Lean tool, work cells can have a significant impact on waste, and it is good to refresh your memory on the 8 Wastes. Many of you reading this playbook already understand the concepts of waste and Lean. For those of you just getting started, these are the 8 Wastes:

Overproduction
Overprocessing
Waiting
Motion
Transportation
Inventory
Defects
Wasted Potential

Overproduction is the act of making more product than necessary and completing it faster than necessary and before it is needed. Overproduced product takes up floor space, requires handling and storage, and could result in potential quality problems if the lot contains defects.

Overprocessing is the practice of extra steps, rechecking, reverifying, and outperforming work. Overprocessing is often conducted in fabrication departments when sanding, deburring, cleaning, or polishing is overperformed. Machines can also overprocess when they are not properly maintained and simply take more time to produce quality parts.

Waiting occurs when important information, tools, and supplies are not readily available, causing machines and people to be idle. Imbalances in workloads and cycle times between processes can also cause waiting.

Motion is the movement of people in and around the work area to look for tools, parts, information, people, and all necessary items that are not available. If a process contains a high level of motion, lead time increases, and the focus on quality begins to decrease. All necessary items should be organized and placed at the point of use so the worker can focus on the work at hand.

Transportation is the movement of parts and product throughout the facility. Often requiring a forklift, hand truck, or pallet jack, transportation exists when consuming processes are far away from each other and are not visible.

Inventory is a waste when manufacturers tie up too much money by holding excessive levels of raw, work-in-process, and finished goods inventory.

Defects are any quality metric that causes rework, scrap, warranty claims, and rework hours from mistakes made in the factory.

Wasted human potential is the act of not properly utilizing employees to the best of their abilities. People are only as successful as the process they are given to work in. If a process inherently has motion, transportation, overprocessing, overproduction, periods of waiting, and defect creation, then that is exactly what involves people. That is wasted human potential.

Cell manufacturing can help you reduce these eight wastes and by doing so will create a much more productive and profitable company for all.

My hope is that you will read this playbook and not only be inspired, but also be able to roll up your sleeves and begin your cell manufacturing journey after the last page is read.

Lean Metrics

To effectively measure your success with cell manufacturing, you need to establish a list of critical shop-floor metrics that can be measured and quantified. On the production floor, these metrics are often called key performance indicators (KPIs). Cell manufacturing is a powerful improvement tool that can have a profound impact on reducing lead times, reducing inventory, increasing output, improving productivity, and affecting many other types of KPIs. In some cases, the change is dramatic. We recommend the following Lean metrics become part of measuring your overall Lean journey.

- Productivity
- Quality
- Inventory
- Floor Space
- Travel Distance
- Throughput Time

Productivity

Productivity is measured in a variety of different ways. Productivity is improved when products are manufactured with less effort. When companies convert production operations from traditional process-based batch manufacturing to leaner work cells, increases to productivity can be significant. I have been part of these types of conversions that have yielded over 90% reductions in lead time while keeping the same number of workers in the work cell. Work cells are designed to operate with a focus on value-added work in almost every movement. So, as workers are focused on value-added work and not the 8 Wastes, productivity can improve to a level never before seen.

Quality

Improvements to quality are more of a secondary benefit of cell manufacturing. In work cells, there is nearly no WIP as single-piece flow is incorporated to provide focus on work. WIP in traditional environments has the potential to hide quality mistakes in higher quantities. Single-piece flow contributes to better focus, and if a mistake is made, it is one unit, not multiple units in a pile of inventory. A batch-processed production line can create higher quantities of mistakes. Standard work procedures and guidelines are developed for work cells to ensure consistency between workers and to help in reducing quality mistakes. You will see as you read through this playbook how cell manufacturing can improve overall internal quality, reduce rework and defects, and ensure a satisfied customer.

Inventory

Inventory is tricky because most companies producing or repairing a product need all three levels of it: raw, WIP, and finished goods. One of the fundamental aspects of cell manufacturing is that the cells create a "sellable" unit. *Sellable* means built, inspected, and packaged so as it leaves the work cell, it is literally ready for delivery. Processes are brought closer and right sized so everything needed to perform the work is done in the work cell, and WIP does not leave the area for any reason. This approach also reduces the level of WIP in the cell and provides better visibility of the work. Another attribute of a work cell is the proper implementation of a Kanban system to ensure parts, material, and supplies are delivered into the work cell as needed to keep the product flowing. I cover this in Chapter 4, and I recommend reading the *Kanban Playbook* from this series.

Kanban systems also help ensure part shortages are kept at a minimum and that excessive inventory is not stored in the work cell.

Floor Space

Floor space comes at a premium, and you need to start looking at the poor use of floor space as hurting the company's ability to grow. Floor space should be used to perform value-added work that creates revenue for the company. It should not be used to store junk or act as a collector of unneeded items. Renting, leasing, or buying a manufacturing building is one of the highest overhead costs. The production floor is in place to serve one purpose: to build products. Although the factory is used for other items, such as holding inventory, shipping, receiving, maintenance, and so on, the production floor should be effectively utilized for value-added work. Value-added work involves the act of building products or the steps needed to change fit, form, or function of the product you intend to sell. Production lines, equipment, and machines all produce a sellable product, and the floor space needed to perform this work should be properly used. Work cell implementations reduce floor space as processes that used to be separated by long distances are brought closer and WIP piles are eliminated.

Buildings cost money, and there is a lot associated with having a facility even if you don't produce product every month. The costs can include the lease, insurance, taxes, utilities, maintenance, and upkeep, so you need to be making money out of it. How much of your space is used to create revenue? Inventory sitting on shelves does not create revenue in a production or repair environment. You can measure your floor space utilization by something called revenue or profit per square foot. As you implement Kanban systems, you will see a reduction in overall carrying costs in your facility. One of the costs is poor use of space, and it is a critical Lean metric.

As a company becomes less organized and unneeded inventory begins to accumulate, more space becomes used for non-value-added items. This creates an increase in waste. Over time, items such as workbenches, garbage cans, chairs, unused equipment, tools, and tables tend to pile up, and valuable production space simply disappears. Rather than reduce inventory and improve floor space use, the general approach is to add. Add building space, racks, shelves, and you want to change your perception of space: better use, fewer non-value-added items, less waste, and less stuff. Work cells are designed for the absolute minimum amount of effort and space to produce product.

Travel Distance

Here is the best way to view travel distance: The farther there is to go, the longer it is going to take. Long production processes can create a lot of waste and can reduce overall performance. Plus, longer-than-needed processes take up floor space. There are two ways to look at travel distance: the distance people walk and the distance inventory (product) is transported.

Travel distance is connected to overall lead times in a process and the entire factory. When WIP is created above required quantities, it takes up valuable floor space and increases the distances that the production line needs. As travel distance increases, floor space becomes improperly used, workers walk farther distances, and lead times are increased. Wait time between processes also increases, and there is added lead time to maneuver inventory.

When work areas are designed incorrectly, they can create a lot of walking for workers, and as they become cluttered, more time is needed to find essential items for work. Work cells take up less space and product travels significantly less than in batch process work areas. Simply put, the less distance product travels, the less time it takes to produce. That is a fact.

Throughput Time

Sometimes used in conjunction with measuring travel distance reduction, throughput time is the time it takes the product to flow in the production process. Throughput time has a direct impact on delivery; the longer it takes product to move through the plant, the longer it takes to be delivered. Of course many variables can extend product lead time, so it is wise to simplify the metric by measuring the time when process 1 grabs raw material to the time it is packaged and ready for shipment. Longer production lines require more workstations, workers, tools, conveyors, supplies, and material, which results in additional cost and WIP as well as extended lead times. A physical reduction in distance equates to less throughput time, allowing an organization to promise more competitive, yet reasonable, delivery dates. Moving inventory through the process faster means less time and money.

Improving these key Lean metrics and using them as a measurement of your success will have a profound impact on the overall financial success and long-term growth of the company. One could look at these Lean metrics simply as process metrics because they can be measured at the shop floor level. Production workers need to work in an efficient environment to be successful contributors to optimal cost, quality, and delivery. Each Lean metric improved complements another, and another, and so on. As you become more experienced as a Lean practitioner, your understanding of how these metrics relate to each other will become second nature.

Chapter 1

Data Collection

Data collection is a critical first step in deciding if your work cell is going to work, look, and maintain high-performance levels. A variety of data collection options are available, and many people can become bogged down with deciding what tool to use. To simplify your focus, I cover the most common and valuable tool: time and motion studies. Time and motion studies are conducted to capture the current state of build time on the products that will be built in your newly implemented work cell. Below is an example of a form I have used for over 10 years.

Work Description:				Time Samples					AVG
Seq #	Work Content	VA	NVA	1	2	3	4	5	

Work Description:					Time Samples				AVG
Seq #	Work Content	VA	NVA	1	2	3	4	5	

Work Description: This section of the form notes the product and product line you are evaluating. There may be different models of the same product that will run through the new work cell, so distinguish between each one.

Seq#: This section is for documenting the number of steps or the sequence number of work.

Work Content: This section documents the work being performed on the product (e.g., assembly, retrieving material, welding, testing, inspection, packaging, etc.). Detailed breakdown of each step is needed.

VA: This stands for value added, which represents the steps that actually change fit, form, or function of the product being assembled. Pieces coming together and material being welded together are good examples. Looking for parts, reading information, and transporting material, although needed, do not add value. Simply check this column if the work content adds value.

NVA: This stands for non–value added, which represents the 8 Wastes of Lean. Using the 8 Wastes as a guideline, simply check the column if it is non–value added and is *not* changing the fit, form, or function of the product.

Time Samples and AVG: It is important not only to capture a time sample for each step but also to collect a few samples to come to an average time that best defines the time needed to complete the task.

Work Description: 3R Electric Bike Series		VA	NVA	Time Samples (Min)					AVG
Seq #	Work Content	VA	NVA	1	2	3	4	5	AVG
1	Attach Brake Cable to Brake Drum and Adjust Tension	X		.55	.65	.50	.62	.65	.59
2	Walk and Retrieve Wire Harness		X	1.1	1.5	1.0	1.0	1.25	1.17
3	Install and Connect Wire Harness	X		.85	.75	.76	.80	.80	.80
4	Secure Wire Harness Wires	X		1.52	1.45	1.40	1.40	1.5	1.45
5	Walk and Retrieve Main Body Sub-assembly		X	2.0	1.90	1.90	2.1	1.85	5.35
6	Install Main Body Sub-assembly and Place in WIP	X		1.52	1.90	1.65	1.85	1.85	1.45
7	Deliver to Hardware Work Area		X	2.55	2.60	2.65	2.6	2.55	2.59
8	Install Black Grommet on Back Suspension	X		.25	.16	.2	.2	.25	.21
9	Walk and Retrieve Hardware from Hardware Storage		X	.66	.68	.7	.7	.65	.68
10	Install (2) 1″ Lock Nuts Over Black Grommet	X		.55	.48	.48	.5	.5	.5
11	Remove Black Grommets from Bag and Place in Bin		X	.25	.25	.25	.2	.24	.24
12	Install Black Grommet on Front Forks	X		.25	.16	.2	.2	.25	.20
13	Install (2) 1″ Lock Nuts Over Black Grommet and Place in WIP	X		.95	.8	.85	.90	.95	.8
14	Deliver to Assembly Area		X	2.1	2.22	2.35	2.2	2.35	2.20
15	Install Seat Post to Main Body Frame	X		1.62	1.90	1.75	1.85	1.85	1.8
16	Walk and Retrieve Seat Cushion		X	.59	.49	.48	.52	.58	.53
17	Install Seat Cushion to Post	X		.66	.68	.7	.7	.65	.68

Continued ■

Work Description: 3R Electric Bike Series				Time Samples (Min)					AVG
Seq #	Work Content	VA	NVA	1	2	3	4	5	
18	Install Seat Quick Release Sub-assembly	X		.25	.25	.25	.2	.24	.24
19	Install Center Panel Lock to Center Panel	X		.95	.8	.76	.80	.80	.82
20	Install Center Panel to Main Body Panel	X		1.63	1.90	1.77	1.85	1.85	1.8
21	Remove Handlebar Frame from Protective Foam			.15	.18	.2	.19	.17	.18
22	Check for Burrs and Deburr as Needed		X	.16	.15	.19	.18	.17	.16
23	Install Handlebar Frame	X		2.35	2.20	2.25	2.35	2.35	2.25
24	Install Quick-Release Sub-assembly and Secure	X		.25	.25	.25	.2	.24	.24
25	Install Rubber Cover on Quick-Release Sub-assembly	X		.15	.19	.2	.15	.17	.17
26	Install Protective Rubber to Seat Post and Place in WIP	X		.68	.68	.71	.71	.65	.68
27	Deliver to Headlight Work Area		X	2.9	2.9	3.1	3.0	2.5	2.8
28	Walk and Retrieve Headlight Sub-assembly from Staging		X	.26	.19	.22	.2	.25	.22
29	Install Headlight Sub-assembly onto Handlebar Frame	X		1.65	1.74	1.7	1.7	1.75	1.7
30	Route Wire Harness Up Handlebar Frame and Secure	X		3.56	3.25	3.5	3.4	3.4	3.4
31	Walk to Retrieve LH Panel from Weld Department		X	3.3	3.0	3.55	3.25	3.4	3.28
32	Walk and Borrow Air Tool		X	.15	.19	.2	.15	.17	.86
33	Install LH Panel	X		.68	.68	.71	.71	.65	.68
34	Stage in WIP Area for Packaging		X	.16	.15	.19	.18	.17	.15

Time Study Totals

Once you have completed time studies on all products that will be incorporated into a new work cell, review the data to pinpoint waste reduction opportunities. This final data will be used as a baseline to compare to future time studies after the work cell is up and running.

Data Categories

- **Total Time:** Add all the work step times (numbers in the AVG column).
- **Value-Added Time:** Add the work step times where the VA box is checked.
- **Non-Value-Added Time:** Add the work step times where the NVA is checked.
- **Calculate the % Waste:** Non-value-added time/total time

Final Data from Time Study Sheet

Total Time: 41.07 minutes
Value-Added Time: 19.78 minutes
Non-Value-Added Time: 21.29 minutes
% Waste: 52%

Now that you have times for all the steps in the process for making the electric bike, let us look at the times in relation to the layout and flow of the process. Look over the time study sheet and you will see Seq.# 7, 14, and 27 are clearly points in the process where the work in progress (WIP) is moved to another area. A visual representation of the process is on Page 8.

Work Description: 3R Electric Bike Series				Time Samples (Min)					AVG
Seq #	Work Content	VA	NVA	1	2	3	4	5	
1	Attach Brake Cable to Brake Drum and Adjust Tension	X		.55	.65	.50	.62	.65	.59
2	Walk and Retrieve Wire Harness		X	1.1	1.5	1.0	1.0	1.25	1.17
3	Install and Connect Wire Harness	X		.85	.75	.76	.80	.80	.80
4	Secure Wire Harness Wires	X		1.52	1.45	1.40	1.40	1.5	1.45
5	Walk and Retrieve Main Body Sub-assembly		X	2.0	1.90	1.90	2.1	1.85	5.35
6	Install Main Body Sub-assembly and Place in WIP	X		1.52	1.90	1.65	1.85	1.85	1.45
7	**Deliver to Hardware Work Area**		X	**2.55**	**2.60**	**2.65**	**2.6**	**2.55**	**2.59**
8	Install Black Grommet on Back Suspension	X		.25	.16	.2	.2	.25	.21
9	Walk and Retrieve Hardware from Hardware Storage		X	.66	.68	.7	.7	.65	.68
10	Install (2) 1″ Lock Nuts Over Black Grommet	X		.55	.48	.48	.5	.5	.5
11	Remove Black Grommets from Bag and Place in Bin		X	.25	.25	.25	.2	.24	.24
12	Install Black Grommet on Front Forks	X		.25	.16	.2	.2	.25	.20
13	Install (2) 1″ Lock Nuts Over Black Grommet and Place in WIP	X		.95	.8	.85	.90	.95	.8
14	**Deliver to Assembly Area**		X	**2.1**	**2.22**	**2.35**	**2.2**	**2.35**	**2.20**
15	Install Seat Post to Main Body Frame	X		1.62	1.90	1.75	1.85	1.85	1.8
16	Walk and Retrieve Seat Cushion		X	.59	.49	.48	.52	.58	.53
17	Install Seat Cushion to Post	X		.66	.68	.7	.7	.65	.68

Continued

Work Description: 3R Electric Bike Series		VA	NVA	Time Samples (Min)					AVG
Seq #	Work Content	VA	NVA	1	2	3	4	5	
18	Install Seat Quick Release Sub-assembly	X		.25	.25	.25	.2	.24	.24
19	Install Center Panel Lock to Center Panel	X		.95	.8	.76	.80	.80	.82
20	Install Center Panel to Main Body Panel	X		1.63	1.90	1.77	1.85	1.85	1.8
21	Remove Handlebar Frame from Protective Foam			.15	.18	.2	.19	.17	.18
22	Check for Burrs and Deburr as Needed		X	.16	.15	.19	.18	.17	.16
23	Install Handlebar Frame	X		2.35	2.20	2.25	2.35	2.35	2.25
24	Install Quick-Release Sub-assembly and Secure	X		.25	.25	.25	.2	.24	.24
25	Install Rubber Cover on Quick-Release Sub-assembly	X		.15	.19	.2	.15	.17	.17
26	Install Protective Rubber to Seat Post and Place in WIP	X		.68	.68	.71	.71	.65	.68
27	**Deliver to Headlight Work Area**		X	2.9	2.9	3.1	3.0	2.5	2.8
28	Walk and Retrieve Headlight Sub-assembly from Staging		X	.26	.19	.22	.2	.25	.22
29	Install Headlight Sub-assembly onto Handlebar Frame	X		1.65	1.74	1.7	1.7	1.75	1.7
30	Route Wire Harness Up Handlebar Frame and Secure	X		3.56	3.25	3.5	3.4	3.4	3.4
31	Walk to Retrieve LH Panel from Weld Department		X	3.3	3.0	3.55	3.25	3.4	3.28
32	Walk and Borrow Air Tool		X	.15	.19	.2	.15	.17	.86
33	Install LH Panel	X		.68	.68	.71	.71	.65	.68
34	Stage in WIP Area for Packaging		X	.16	.15	.19	.18	.17	.15

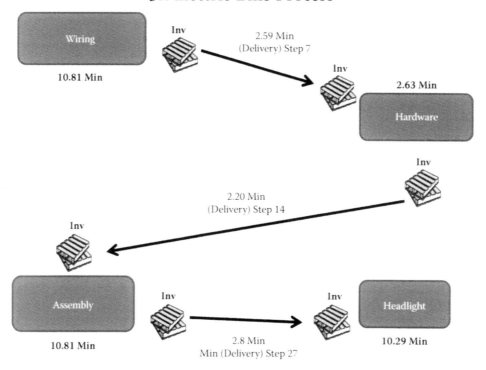

3R Electric Bike Process

The final pieces of data that help calculate a final overall lead time are the amount of time the WIP sits after and before each process. This queue time is often the largest contributor to overall lead time as the time it takes the pile to move is based on how long the operator is allowed to pile up inventory.

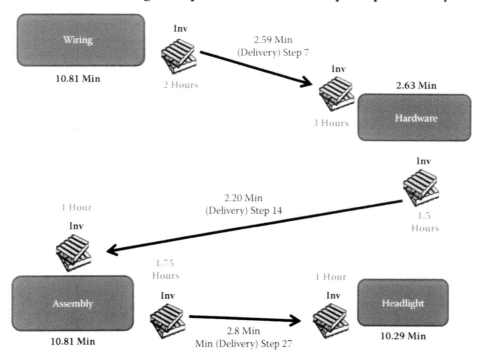

Lead Time and Final Numbers

Total Working Build Time and Transportation Time

Total Time: 41.07 minutes
Value-Added Time: 19.78 minutes
Non-Value-Added Time: 21.29 minutes
% Waste: 52%
Delays Due to WIP: 10.25 hours
Total Lead Time of 3R Electric Bike: 10 hours and 56 minutes

Do not forget, value added is only 19.78 minutes of the overall lead time.

You could dig deeper into the current state of the process. Deeper data could include

- Travel distance from spaghetti diagram (next section)
- Percentage downtime on equipment
- Percentage rework or defects
- Number of workers
- Quantity of WIP units
- Floor space use

But, for the purposes of this playbook, the time study information and general layout are enough to illustrate how to move from traditional process-based production to cell manufacturing.

Spaghetti Diagram

Spaghetti diagrams provide insight on where required stations, equipment, material, and all items are currently located and need to be brought closer to the work cell.

- Visual representation of the work area
- Computer-aided design (CAD) or hand drawing
- Illustrates where people and material are and walking distance
- Captures current state:
  - Travel distance (feet)
  - Time in travel (minutes)
  - Labor (dollars)

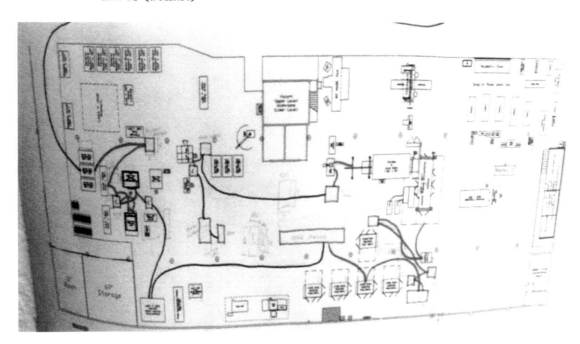

Families

It is important to capture time studies on as many product lines as will be incorporated into the new work cells. While conducting your analysis, make sure to separate the product lines into families of commonality. Not all products can be built on the same work cell or assembly line. Often, organizations build everything on one production line; an assessment of build commonality was never performed. Work cells work best when the products being produced are grouped into families, and it is these families that are used to determine the number of work cells needed.

Your time study information and spaghetti diagrams will help you see commonality in work and required workstations/equipment. Each family will have a certain number of product lines; these families are then used to design suitable work cells.

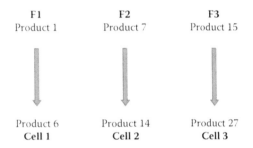

Simple time and motion studies can provide plenty of data on the current state of any process. Many people become caught up in wanting to collect more information over an extended period of time to become more confident in the integrity of the data. It has been my experience in this field that the results will not be drastically different, and the direction you would take with more data would lead in the same direction.

Now that you have your baseline time studies, you are ready to start designing the work cells.

Chapter 2

Work Cell Line Balancing

Introduction

A workload balancing exercise helps ensure a smooth-running work cell or assembly process. Workload balancing is an evaluation of the work content in the time study sheet to distribute work among the required number of workers. Certain calculations are needed to conduct workload balancing; these are covered in this chapter. Topics covered are

- Takt time
- Available time
- Workload balancing

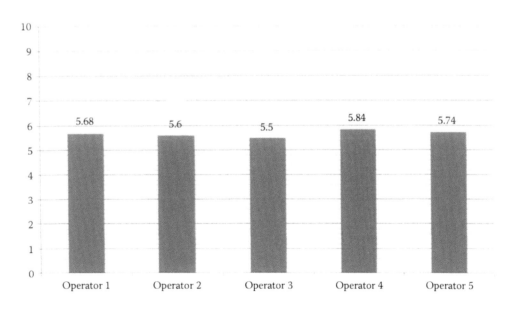

Takt Time

Takt time is defined as the product completion time of a work cell or assembly process. *Takt* is a German word meaning "pulse" or "rhythm." To simplify the definition, it is how often a completed product must be produced to keep up with customer demand. Takt time is used to properly balance the workloads between operators in the work cell.

$$\text{Takt Time} = \frac{\text{Available Time (Number of Shifts)}}{\text{Daily Demand}}$$

Available time is the time available to work in the work cell minus startup time, break times, and cleanup time each day.

Available Time Example

Clock in clock time (in the building):	8 hours and 30 minutes
Startup time:	15 minutes
First break:	10 minutes
Lunch break:	30 minutes
Second break:	10 minutes
End-of-day cleanup:	15 minutes
Total time not working:	80 minutes *or* 1 hour and 20 minutes
Available time:	7 hours and 10 minutes

The number of shifts is optional, and it depends on whether you currently have more than one shift or would like to consider running the work cell for two shifts, for example. I'll show a few options. Also, daily demand is essentially the design rate at which the work cell will work. It does not mean you are stuck to that volume; there is flexibility in a work cell. I will show those options as well.

Example 2.1

Available time: 7 hours and 10 minutes *or* 430 minutes
Shifts: 1
Daily demand: 50 units
Takt time = 430(1)/50 = 8.6 minutes

Example 2.2

Available time: 7 hours and 10 minutes *or* 430 minutes
Shifts: 2
Daily demand: 50 units
Takt time = 430(2)/50 = 17.2 minutes

In this example, with a second shift, you do not need the product completed as often as you are utilizing time in the evening. You will essentially make 25 units on each shift.

Example 2.3

Available time: 7 hours and 10 minutes *or* 430 minutes
Shifts: 1
Daily demand: 30 pallets
Takt time = 430(1)/30 = 14.3 minutes

In Example 2.3, the work cell is tracking output based on the number of pallets not units. In many cases, measuring units is not the proper approach. In automated processes, as examples at a high rate, bags may be filled or wood may be cut. In this case, the pallet really is the finished product because the customer purchases the product in pallets not units. You can get creative on how you calculate Takt time; just make sure it is an increment that represents a "sellable" unit.

Do not forget that you must look at all the products to be made in the work cell and establish how many of each one will be needed. For instance

Model	Quantity
1A	5
2R	6
2T	11
3G	4
Total daily demand	26

Takt time = 430(1)/26 = 16.5 minutes. You are able to build more or less of each model while retaining a total of 26 for the day.

Workload Balancing

Once you have Takt time calculated, you can start workload balancing. Use the time study sheet created in Chapter 1 as a guideline. When performing the workload balancing portion of your work cell design, it is good to follow three basic rules:

1. Remove non-value-added work
2. Balance by Takt time
3. Balance by work

Remove Non-Value-Added Work

Take out the time study sheet and highlight where the non-value-added work is identified. You can safely make some assumptions about your future work cell design by determining which steps will be eliminated by cell manufacturing. Any step that identifies walking, retrieving, and delivering will be eliminated from the future work content. Once you have highlighted the steps, create a revision of the time study sheet with the non-value-added steps deleted from the sheet. Page 17 shows the highlighted steps, and page 19 shows you the steps deleted. I discuss each removed step on page 20.

Work Description: 3R Electric Bike Series				Time Samples (Min)					AVG
Seq #	Work Content	VA	NVA	1	2	3	4	5	
1	Attach Brake Cable to Brake Drum and Adjust Tension	X		.55	.65	.50	.62	.65	.59
2	Walk and Retrieve Wire Harness		X	1.1	1.5	1.0	1.0	1.25	1.17
3	Install and Connect Wire Harness	X		.85	.75	.76	.80	.80	.80
4	Secure Wire Harness Wires	X		1.52	1.45	1.40	1.40	1.5	1.45
5	Walk and Retrieve Main Body Sub-assembly		X	2.0	1.90	1.90	2.1	1.85	5.35
6	Install Main Body Sub-assembly and Place in WIP	X		1.52	1.90	1.65	1.85	1.85	1.45
7	Deliver to Hardware Work Area		X	2.55	2.60	2.65	2.6	2.55	2.59
8	Install Black Grommet on Back Suspension	X		.25	.16	.2	.2	.25	.21
9	Walk and Retrieve Hardware from Hardware Storage		X	.66	.68	.7	.7	.65	.68
10	Install (2) 1″ Lock Nuts Over Black Grommet	X		.55	.48	.48	.5	.5	.5
11	Remove Black Grommets from Bag and Place in Bin		X	.25	.25	.25	.2	.24	.24
12	Install Black Grommet on Front Forks	X		.25	.16	.2	.2	.25	.20
13	Install (2) 1″ Lock Nuts Over Black Grommet and Place in WIP	X		.95	.8	.85	.90	.95	.8
14	Deliver to Assembly Area		X	2.1	2.22	2.35	2.2	2.35	2.20
15	Install Seat Post to Main Body Frame	X		1.62	1.90	1.75	1.85	1.85	1.8
16	Walk and Retrieve Seat Cushion		X	.59	.49	.48	.52	.58	.53
17	Install Seat Cushion to Post	X		.66	.68	.7	.7	.65	.68
18	Install Seat Quick Release Sub-assembly	X		.25	.25	.25	.2	.24	.24

Continued

Work Description: 3R Electric Bike Series				Time Samples (Min)					AVG
Seq #	Work Content	VA	NVA	1	2	3	4	5	
19	Install Center Panel Lock to Center Panel	X		.95	.8	.76	.80	.80	.82
20	Install Center Panel to Main Body Panel	X		1.63	1.90	1.77	1.85	1.85	1.8
21	Remove Handlebar Frame from Protective Foam			.15	.18	.2	.19	.17	.18
22	Check for Burrs and Deburr as Needed		X	.16	.15	.19	.18	.17	.16
23	Install Handlebar Frame	X		2.35	2.20	2.25	2.35	2.35	2.25
24	Install Quick-Release Sub-assembly and Secure	X		.25	.25	.25	.2	.24	.24
25	Install Rubber Cover on Quick-Release Sub-assembly	X		.15	.19	.2	.15	.17	.17
26	Install Protective Rubber to Seat Post and Place in WIP	X		.68	.68	.71	.71	.65	.68
27	Deliver to Headlight Work Area		X	2.9	2.9	3.1	3.0	2.5	2.8
28	Walk and Retrieve Headlight Sub-assembly from Staging		X	.26	.19	.22	.2	.25	.22
29	Install Headlight Sub-assembly onto Handlebar Frame	X		1.65	1.74	1.7	1.7	1.75	1.7
30	Route Wire Harness Up Handlebar Frame and Secure	X		3.56	3.25	3.5	3.4	3.4	3.4
31	Walk to Retrieve LH Panel from Weld Department		X	3.3	3.0	3.55	3.25	3.4	3.28
32	Walk and Borrow Air Tool		X	.15	.19	.2	.15	.17	.86
33	Install LH Panel	X		.68	.68	.71	.71	.65	.68
34	Stage in WIP Area for Packaging		X	.16	.15	.19	.18	.17	.15

As you can see below, there are now 21 steps remaining from the original 34. These steps will not be needed in a cell manufacturing environment. It streamlines the work down to the value-added steps. There may be small amounts of non-value-added work but for the purposes of our work cell, most of it has been removed. See page 20 for more explanation.

Work Description: 3R Electric Bike Series		VA	NVA	Time Samples (Min)					AVG
Seq #	Work Content			1	2	3	4	5	
1	Attach Brake Cable to Brake Drum and Adjust Tension	X		.55	.65	.50	.62	.65	.59
2	Install and Connect Wire Harness	X		.85	.75	.76	.80	.80	.80
3	Secure Wire Harness Wires	X		1.52	1.45	1.40	1.40	1.5	1.45
4	Install Main Body Sub-assembly and Place in WIP	X		1.52	1.90	1.65	1.85	1.85	1.45
5	Install Black Grommet on Back Suspension	X		.25	.16	.2	.2	.25	.21
6	Install (2) 1″ Lock Nuts Over Black Grommet	X		.55	.48	.48	.5	.5	.5
7	Install Black Grommet on Front Forks	X		.25	.16	.2	.2	.25	.20
8	Install (2) 1″ Lock Nuts Over Black Grommet and Place in WIP	X		.95	.8	.85	.90	.95	.8
9	Install Seat Post to Main Body Frame	X		1.62	1.90	1.75	1.85	1.85	1.8
10	Install Seat Cushion to Post	X		.66	.68	.7	.7	.65	.68
11	Install Seat Quick Release Sub-assembly	X		.25	.25	.25	.2	.24	.24
12	Install Center Panel Lock to Center Panel	X		.95	.8	.76	.80	.80	.82
13	Install Center Panel to Main Body Panel	X		1.63	1.90	1.77	1.85	1.85	1.8
14	Remove Handlebar Frame from Protective Foam			.15	.18	.2	.19	.17	.18
15	Install Handlebar Frame	X		2.35	2.20	2.25	2.35	2.35	2.25
16	Install Quick-Release Sub-assembly and Secure	X		.25	.25	.25	.2	.24	.24
17	Install Rubber Cover on Quick-Release Sub-assembly	X		.15	.19	.2	.15	.17	.17
18	Install Protective Rubber to Seat Post and Place in WIP	X		.68	.68	.71	.71	.65	.68

Continued

Work Description: 3R Electric Bike Series			VA	NVA	Time Samples (Min)					AVG
Seq #	Work Content		VA	NVA	1	2	3	4	5	
19	Install Headlight Sub-assembly onto Handlebar Frame		X		1.65	1.74	1.7	1.7	1.75	1.7
20	Route Wire Harness Up Handlebar Frame and Secure		X		3.56	3.25	3.5	3.4	3.4	3.4
21	Install LH Panel		X		.68	.68	.71	.71	.65	.68

Workload Balancing

Non-Value-Added Work Content	Elimination
Walk and retrieve wire harness	Stored in work cell
Walk and retrieve main body sub-assembly	Main body sub-assembly delivered (Kanban)
Deliver to hardware work area	Hardware station to be in work cell
Walk and retrieve hardware from hardware storage	Hardware stored in work cell
Remove black grommets from bag and place in bin	Removed by material handlers
Deliver to assembly area	Assembly station to be in work cell
Walk and retrieve seat cushion	Stored in work cell
Check for burrs and deburr as needed	Performed where I should (fabrication)
Deliver to headlight work area	Headlight station to be in work cell
Walk and retrieve headlight sub-assembly from staging	Headlight sub-assembly delivered (Kanban)
Walk to retrieve LH panel from weld department	LH panel delivered (Kanban)
Walk and borrow air tool	Tools at point of use in work cell
Stage in work-in-progress (WIP) area for packaging	Unchanged or a packaging station can be incorporated later into the work cell

Some of the other Lean concepts that are applied to these wastes are reviewed in further chapters.

Balance by Takt Time

This balancing step is the easiest as it involves adding blocks of work from the time study information so it equals Takt time. For the 3R electric bike work cell, I will use a daily demand of 90 bikes.

430 minutes (1 Shift)/90 Bikes = Takt time of **4.77 minutes**

Work Description: 3R Electric Bike Series			AVG	
Seq #	Work Content	VA		
1	Attach Brake Cable to Brake Drum and Adjust Tension	X	.59	
2	Install and Connect Wire Harness	X	.80	**4.5 minutes**
3	Secure Wire Harness Wires	X	1.45	
4	Install Main Body Sub-assembly and Place in WIP	X	1.45	
5	Install Black Grommet on Back Suspension	X	.21	
6	Install (2) 1″ Lock Nuts Over Black Grommet	X	.5	
7	Install Black Grommet on Front Forks	X	.20	
8	Install (2) 1″ Lock Nuts Over Black Grommet and Place in WIP	X	.8	**4.22 minutes**
9	Install Seat Post to Main Body Frame	X	1.8	
10	Install Seat Cushion to Post	X	.68	
11	Install Seat Quick Release Sub-assembly	X	.24	
12	Install Center Panel Lock to Center Panel	X	.82	
13	Install Center Panel to Main Body Panel	X	1.8	**5.05 minutes**
14	Remove Handlebar Frame from Protective Foam		.18	
15	Install Handlebar Frame	X	2.25	
16	Install Quick-Release Sub-assembly and Secure	X	.24	
17	Install Rubber Cover on Quick-Release Sub-assembly	X	.17	
18	Install Protective Rubber to Seat Post and Place in WIP	X	.68	**2.79 minutes**
19	Install Headlight Sub-assembly onto Handlebar Frame	X	1.7	
20	Route Wire Harness Up Handlebar Frame and Secure	X	3.4	**4.08 minutes**
21	Install LH Panel	X	.68	

Balance by Work

Our work content is almost balanced but we need to shift work around a bit throughout the build sequence to further balance to our 4.77 minute Takt time. Depending on the complexity of the work, you can often shift work up and down the sheet to further balance the process. Move Step 13 between Step 15 and Step 16. Then move Step 18 to right after Step 15. Process more balanced.

Seq #	Work Content	VA	AVG	
	Work Description: 3R Electric Bike Series		*AVG*	
6	Install (2) 1″ Lock Nuts Over Black Grommet	X	.5	
7	Install Black Grommet on Front Forks	X	.20	
8	Install (2) 1″ Lock Nuts Over Black Grommet and Place in WIP	X	.8	**4.22 minutes**
9	Install Seat Post to Main Body Frame	X	1.8	
10	Install Seat Cushion to Post	X	.68	
11	Install Seat Quick Release Sub-assembly	X	.24	
12	Install Center Panel Lock to Center Panel	X	.82	
14	Remove Handlebar Frame from Protective Foam		.18	**3.93 minutes**
15	Install Handlebar Frame	X	2.25	
18	Install Protective Rubber to Seat Post and Place in WIP	X	.68	
13	Install Center Panel to Main Body Panel	X	1.8	
16	Install Quick-Release Sub-assembly and Secure	X	.24	
17	Install Rubber Cover on Quick-Release Sub-assembly	X	.17	**3.91 minutes**
19	Install Headlight Sub-assembly onto Handlebar Frame	X	1.7	

The 3R electric bike work cell is much more balanced. It is not perfect, but at least it is close enough to ensure consistent flow throughout the work cell once it is up and running. Step 13 is not valued as removing parts from any form of packaging is waste, but we leave it there, assuming the protective foam needs to stay on the part until installation.

Seq #	*Work Content*	*VA*	*AVG*	
	*Work Description: 3R Electric Bike Series*			
1	Attach Brake Cable to Brake Drum and Adjust Tension	X	.59	
2	Install and Connect Wire Harness	X	.80	
3	Secure Wire Harness Wires	X	1.45	**4.5 minutes**
4	Install Main Body Sub-assembly and Place in WIP	X	1.45	
5	Install Black Grommet on Back Suspension	X	.21	
6	Install (2) 1″ Lock Nuts Over Black Grommet	X	.5	
7	Install Black Grommet on Front Forks	X	.20	
8	Install (2) 1″ Lock Nuts Over Black Grommet and Place in WIP	X	.8	**4.22 minutes**
9	Install Seat Post to Main Body Frame	X	1.8	
10	Install Seat Cushion to Post	X	.68	
11	Install Seat Quick Release Sub-assembly	X	.24	
12	Install Center Panel Lock to Center Panel	X	.82	
13	Remove Handlebar Frame from Protective Foam		.18	
14	Install Handlebar Frame	X	2.25	**3.93 minutes**
15	Install Protective Rubber to Seat Post and Place in WIP	X	.68	
16	Install Center Panel to Main Body Panel	X	1.8	
17	Install Quick-Release Sub-assembly and Secure	X	.24	
18	Install Rubber Cover on Quick-Release Sub-assembly	X	.17	**3.91 minutes**
19	Install Headlight Sub-assembly onto Handlebar Frame	X	1.7	
20	Route Wire Harness Up Handlebar Frame and Secure	X	3.4	**4.08 minutes**
21	Install LH Panel	X	.68	

Chapter 3

Work Cell Line Design

Introduction

Once you have completed the line-balancing exercises, you have one more step to complete prior to implementation. You need to spend some time designing the work cell and ensuring you have identified all the required items for implementation. In this chapter, I refer to the 3R electric bike work cell to illustrate the steps, then I leave that example behind and focus on other examples. The other cell manufacturing design examples are from real line designs. Now, let us see what the data look like when the five individual workstations are broken out.

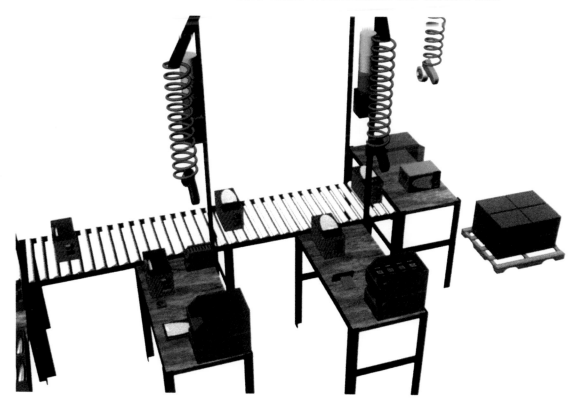

Work Description: 3R Electric Bike Series			St. 1
Seq #	Work Content	VA	
1	Attach Brake Cable to Brake Drum and Adjust Tension	X	.59
2	Install and Connect Wire Harness	X	.80
3	Secure Wire Harness Wires	X	1.45
4	Install Main Body Sub-assembly and Place in WIP	X	1.45
5	Install Black Grommet on Back Suspension	X	.21

4.5 minutes

Work Description: 3R Electric Bike Series			St. 2
Seq #	Work Content	VA	
6	Install (2) 1″ Lock Nuts Over Black Grommet	X	.5
7	Install Black Grommet on Front Forks	X	.20
8	Install (2) 1″ Lock Nuts Over Black Grommet and Place in WIP	X	.8
9	Install Seat Post to Main Body Frame	X	1.8
10	Install Seat Cushion to Post	X	.68
11	Install Seat Quick Release Sub-assembly	X	.24

4.22 minutes

Work Description: 3R Electric Bike Series			St. 3
Seq #	Work Content	VA	
12	Install Center Panel Lock to Center Panel	X	.82
13	Remove Handlebar Frame from Protective Foam		.18
14	Install Handlebar Frame	X	2.25
15	Install Protective Rubber to Seat Post and Place in WIP	X	.68

3.93 minutes

Work Description: 3R Electric Bike Series			St. 4
Seq #	*Work Content*	*VA*	
16	Install Center Panel to Main Body Panel	X	1.8
17	Install Quick-Release Sub-assembly and Secure	X	.24
18	Install Rubber Cover on Quick-Release Sub-assembly	X	.17
19	Install Headlight Sub-assembly onto Handlebar Frame	X	1.7

3.91 minutes

Work Description: 3R Electric Bike Series			St. 5
Seq #	*Work Content*	*VA*	
20	Route Wire Harness Up Handlebar Frame and Secure	X	3.4
21	Install LH Panel	X	.68

4.08 minutes

Breaking out the work content in this manner helps you visualize the work each operator will do and where they will be in the work cell. Although they are assigned specific tasks, they do work as a team; this team is called a *work team*. Their sole purpose as a work team is to complete the product running through the work cell.

Using the work content identified in the line-balancing sheets, simply create a list of everything that is needed in that workstation. This list is used during implementation to ensure the cell has everything it needs to run properly. Whether you are creating a new work area or converting an existing process, this list is the blueprint for implementation. I use station 4 as the example.

Work Description: 3R Electric Bike Series			St. 4
Seq #	Work Content	VA	
16	Install Center Panel to Main Body Panel	X	1.8
17	Install Quick-Release Sub-assembly and Secure	X	.24
18	Install Rubber Cover on Quick-Release Sub-assembly	X	.17
19	Install Headlight Sub-assembly onto Handlebar Frame	X	1.7

Tools
  Air ratchet
  ¼" socket
  Flat-head screwdriver

Equipment
  Headlight tester

Parts

Part Description	Part Number
Center panel	45-883-21
Quick-release Sub-assembly	N/A
Rubber cover	55-8910-11
¼" nuts	10-2111-10
¼" Tek screws	10-8792-10

The following list is to ensure the workstation has the required items for building the 3R bike. That list will help you identify how tools, equipment, and parts will be stored in the workstations. The final list that follows provides a guideline for constructing each workstation and the work cell itself.

Work Description: 3R Electric Bike Series			St. 4
Seq #	Work Content	VA	
16	Install Center Panel to Main Body Panel	X	1.8
17	Install Quick-Release Sub-assembly and Secure	X	.24
18	Install Rubber Cover on Quick-Release Sub-assembly	X	.17
19	Install Headlight Sub-assembly onto Handlebar Frame	X	1.7

Station 4
    Lift table and lazy Susan
    Compressed air lines
    Electrical lines
    Station light
    Communication lift
    Tool retractor

Items to Consider

Antifatigue mats

Turntables

Mobile work cells:
"train"-style movement

Communication
lights

Lift tables

Rollers for moving in
and out of stations

Tool retractors

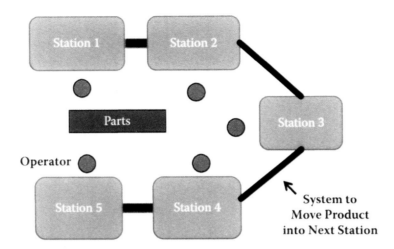

In our 3R Electric Bike example, each person is assigned to a workstation based on the work content established in the load balancing exercise. Everything they need is right at point of use: parts, tools, material, sub-assembly from other areas, etc. Ideally we would incorporate packaging as well as a final station. Once you have done that, then the product is truly being completed in the work cell. Plus make sure the work is balanced to your 4.77 minute Takt time.

I realize this is a general example, so let's show you real life work cells in both 3D models and pictures.

Results

Measure	Before	After
Lead time	10 hours 56 minutes	20.38 minutes
Work time	41.07 minutes	20.38 minutes
Non-value-added time	21.29 minutes	.18 minutes
Percentage waste	52%	Less than 1%

Ruler making work cell

- U-shaped cell
- Point-of-use material
- Takt time: 13 seconds

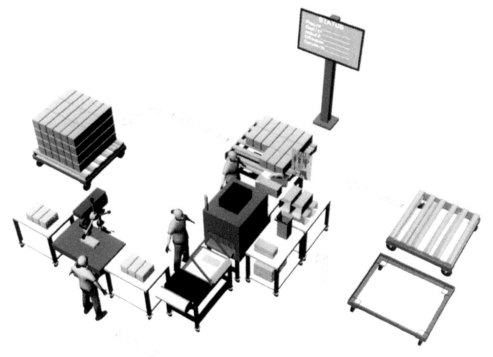

Ruler making work cell

- Lift table at end of cell
- Rulers go directly in box
- Box sits on pallet on lift table
- Value added to value added

- Pallet cart with empty pallet
- Cart rolls over lift table
- Lift table raises to waist height
- Once full, lower and roll off
- No pallet racks

Full pallet rolls into pallet station

Final product staged in Finished Goods area

Results

Measure	Before	After
Lead time	7 hours	84 seconds
Output	1150 rulers	1938 rulers

Standard cost per unit reduction: 18%

Bottle filling work cell

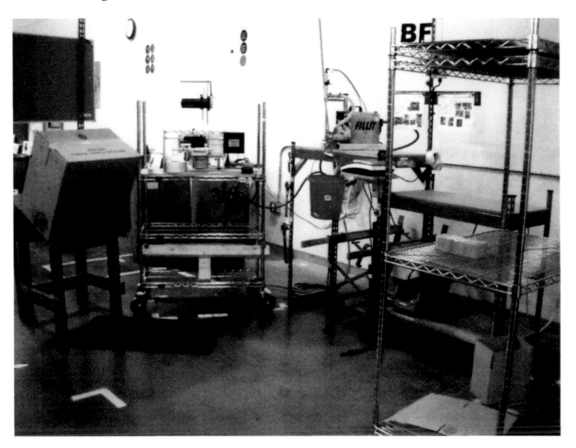

Single-piece flow: Label, fill, cap, box, done

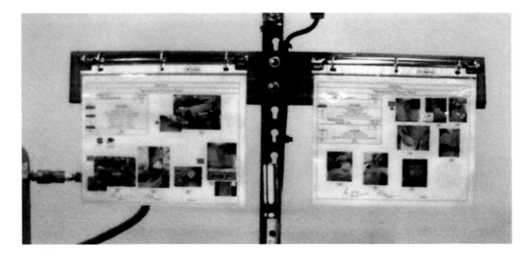

Visual work instructions in station

Parts in front of workbench

Work-in-progress (WIP) staging
in-between stations

Stations signs

Good upfront planning and designing of a work cell will set the tone for successful implementation. As you prepare for implementation keep in mind that there will be some resistance by the production workers and even supervisors. It is a completely different way of working. There is a higher level of urgency, workers must follow the standard work established in each workstation, and when problems arise, there is not added WIP to act a buffer. You must deal with these issues head on.

With that said, you can reduce this transition by proper planning; this chapter showed you those required items. Now, it is time to put the plan into place.

Chapter 4

Implementation

Introduction

In the 15 years I have been a Lean practitioner, both when I worked for companies and in the 11 years of running Kaizen Assembly, I have always been a proponent of kaizen events. Kaizen events are intended to be the mechanism for Lean implementation. This chapter walks you through a 4-day kaizen event to illustrate how to implement your work cell. There are other approaches to implementation, but kaizen events have proven to be the most effective for me.

Here are the key attributes of a kaizen event:

- Select a team leader
- Select four to six team members
- Schedule the event: 4 days in a row

Make sure the team is 100% dedicated to the 4 days and that their normal day-to-day work is given to someone else in their absence. There is a lot of work to be done, and their involvement needs to be uninterrupted. Also, make sure all the data, supplies, and work cell items discussed in Chapter 3 are ready.

Day 1: Sort

Day 1 starts with removing all unneeded items from the "old" production process that is being converted to a work cell. Try to get down to the minimum required essentials. Use a Red Tag process to properly identify and remove items. Information on this tag will help management make decisions on how to dispose of each Red-Tagged item. Place these sorted items into a temporary staging area. Once sorting is complete, you will have a nice visual of what is left and truly needed for the work cell—stay to the design.

Red Tags

Work cell preparation Red Tag area

Red Tag items from old production line: not needed in new work cell

Red Tag area: rented container

Day 2: Line Setup and Dry Run

Most of Day 1 will be sorting and reviewing final details of the work cell design with the team. Day 2 is dedicated to putting the workstations, equipment, and all essential components in place. Do not worry about tools and parts just yet. First, place the items that take up floor space. You will go into detail on Days 3 and 4 as the other part of Day 2 is to conduct a dry run to verify flow, time, and workload balancing.

Day 2 of implementation: main floor items in place for dry run

Lift table going into place

Workbenches going into place

(see above)

A dry run is intended to verify the work cell setup, the time studies, and the line-balancing data. Place people in their positions and hand out the new standard work for each station. Make sure they have tools and enough parts to do about a 1-hour dry run. This dry run is led by the team leader, and he or she acts as a conductor to guide the workers. This part will require a little babysitting at first but once the workers get into a rhythm, the work cell will start to operate as designed. Once the work cell is producing product at the designed Takt time, you can move on to the 5S activities, tool placement, and material presentation.

Days 3–4: 5S, Tools, Parts, and Material

Once you feel the work cell is operating at Takt time consistently, it is time to fine-tune and verify the workstations have everything they need to meet that Takt time. This chapter provides images of how to present tools, parts, and material to the work cell. Although there are examples of 5S, shadow boards, and visual material, I do not explain 5S in depth. However, the pictures and explanations will do you well in understanding the activities of Days 3 and 4.

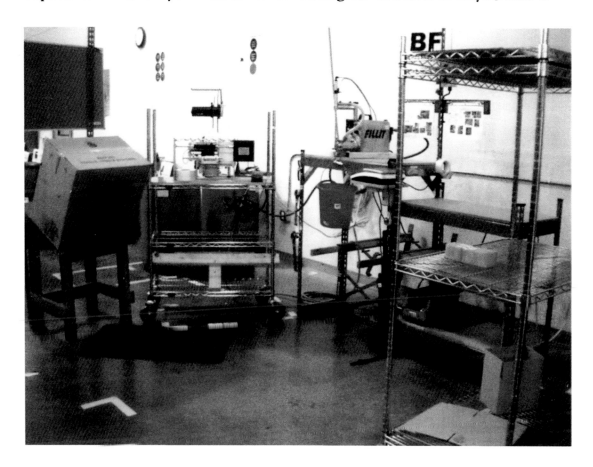

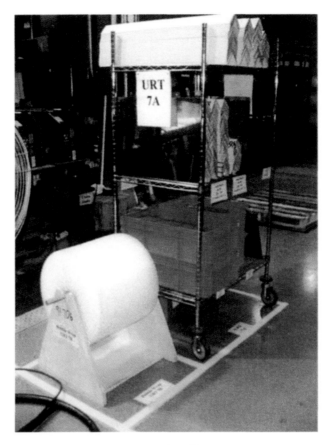

Mark off the work area and items on the floor.

- This is the address for items.
- Items in rack are labeled URT 7A.

Address   Floor markings

Packaging cart and matching labels

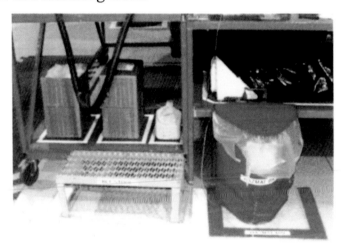

Stool and hazmat bin

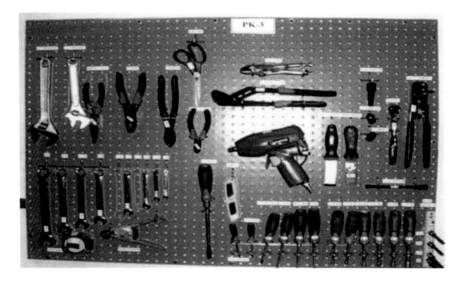

Shadow board in workstation

■ Shadow board
   on machine
■ Swivels as needed
   for operator

■ Workbench
■ Items marked off
   and labeled

- Boxes in packaging station
- Kanban card system for reordering

- Packaging station.
- Everything has a home.

- Part presentation in work cell

■ Floor locations
  for garbage and table
■ Work area marked off

■ Incoming staging for material
■ Directional tape
■ Location for cart and label

Material handlers are a vital component of any work cell. Generally deployed into production areas, material handlers keep material flowing in and out of the work cell, allowing the workers to focus on performing value-added work. Many companies view material handlers as a waste of company resources as their work is considered indirect. Direct labor creates value for the company by producing product that can be sold to bring in money. Material handlers are considered indirect, and their contributions only take money from the company. Although there is truth to these statements as far as direct and indirect, allow me to paint you a different picture.

Imagine a race team: There is a driver and a pit crew. There are other people/employees of a race team, such as those in management, but let us stay with the driver and pit crew. The driver can be looked at as direct labor. His or her job is to race the cart and complete the required number of laps in the least amount time compared to the competition. The act of racing is value added. Imagine if the driver had to get out of his or her car and perform each pit stop: change tires, fill the gas tank, clean the window, and so on. Does this seem silly? The race is lost. The pit crew acts as the material handlers of the "system," and their contributions are valuable. Also, if you were to place three videos of three pit stops from one race and place them on a screen at the same time, their movements and time would be virtually identical. There is structure to what they do, where they go, each step taken, and more.

Kanban Card or Bin Drop Point

- Operators place empty bin or card at the drop point.
- Operators turn on the light to signal material handlers.
- Material handlers come to the drop point.
- Material handlers retrieve cards and bins.
- Material handlers turn light off.
- Material handlers return with parts material.

Drop points are needed for bins and cards and can be strategically placed throughout the work area; each operator on the production line can be assigned to a drop point. By assigning operators to drop points, you keep the flow of material in and out even. This ensures you do not clutter drop points and do not overload material handlers.

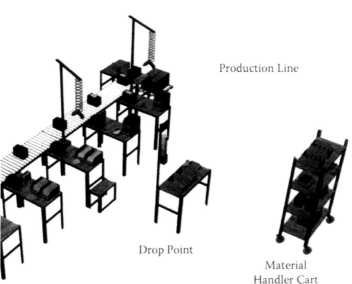

Production Line

Drop Point

Material
Handler Cart

In many cases, the progress and flow of the line can be gauged by the flow of the material-handling system. Because the quantities of parts in the work area are based on a rotation with minimum and maximum quantities, parts should be flowing in and out based on output. If there are any issues with output, signals for replenishment will not appear as planned.

Communication Lights

- Lights act as communication to everyone.
- Lights are great for communication to material handling.
- Lights are a signal communicating empty bins or cards in a workstation.

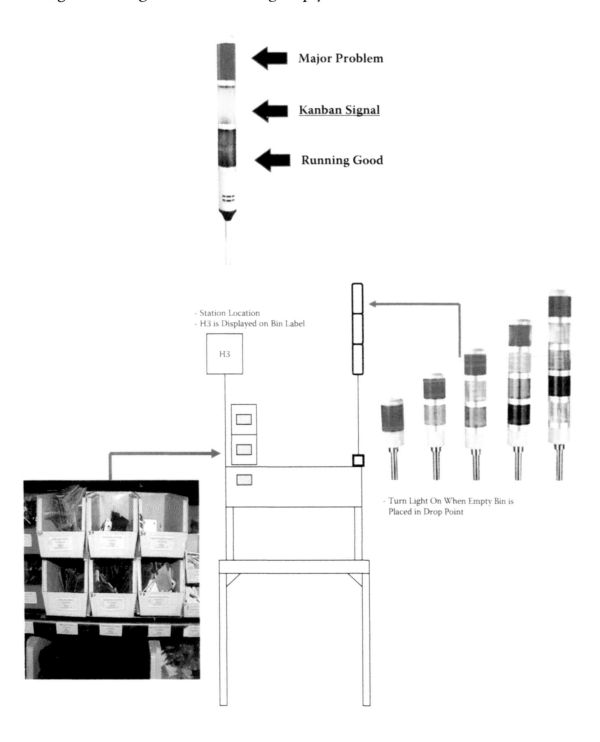

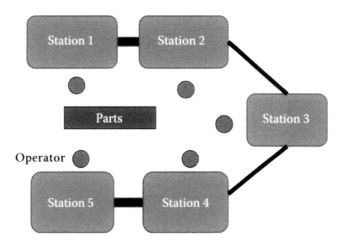

Some work cells will take longer than 4 days to implement. As the required workstations increase, so does the number of tools, carts, benches, material, parts, and so on. As you implement work cells more often, you will become better prepared for determining the length of time needed.

The day after implementation, it is recommended that you and all the required supervisors and managers spend time supporting the process and work out any issues that may arise as you now start "full" production. As mentioned, expect a level of resistance and transition. My best advice is to be patient and work the new process. If the data collection and line-balancing work is correct, you should have no problem achieving daily consistency in the output of the work cell.

Chapter 5

Work Instructions and Production Control Boards

Lean processes need the proper documentation to illustrate how the product is made as it moves from person to person. Most work instructions I have come across are hard to read, contain the wrong information, contain too many words, and are not visual. Chapter 5 covers how to create visual work procedures, provides real-life examples, and gives ways to install them at point of use.

- Make them workstation/job specific
- Install at point of use
- Make them easy to understand
- Be creative and use shapes, arrows, boxes, and so on
- Use few words
- Use symbols and icons to illustrate action
- Use computer-aided design (CAD) drawings or pictures

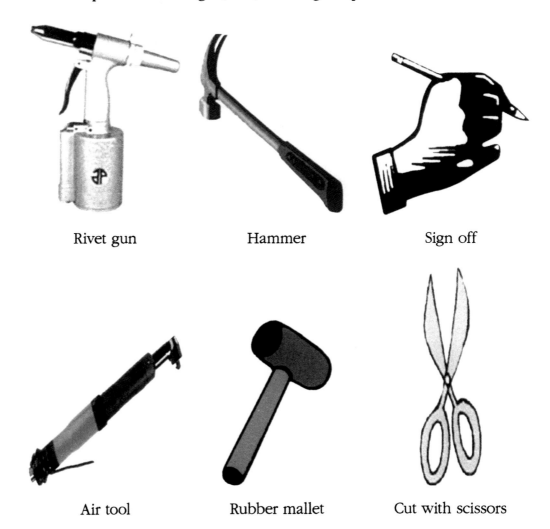

Rivet gun	Hammer	Sign off
Air tool	Rubber mallet	Cut with scissors

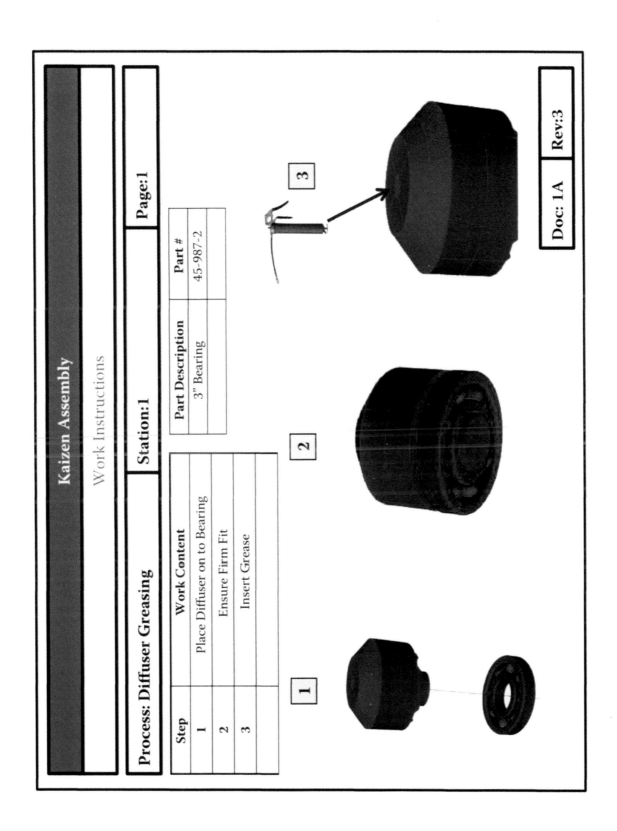

Kaizen Assembly

Work Instructions

Process: Diffuser Greasing	Station:1	Page:1

Step	Work Content
1	Place Diffuser on to Bearing
2	Ensure Firm Fit
3	Insert Grease

Part Description	Part #
3" Bearing	45-987-2

Doc: 1A	Rev:3

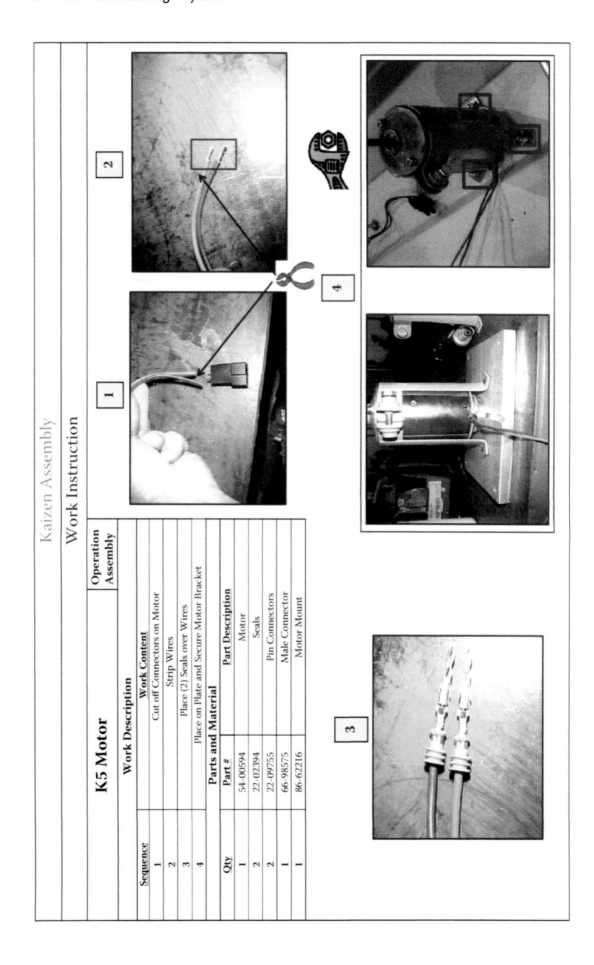

Kaizen Assembly

Work Instruction

K5 Motor	Operation Assembly

Work Description

Sequence	Work Content
1	Cut off Connectors on Motor
2	Strip Wires
3	Place (2) Seals over Wires
4	Place on Plate and Secure Motor Bracket

Parts and Material

Qty	Part #	Part Description
1	54-00594	Motor
2	22-02394	Seals
2	22-09755	Pin Connectors
1	66-98575	Male Connector
1	86-62216	Motor Mount

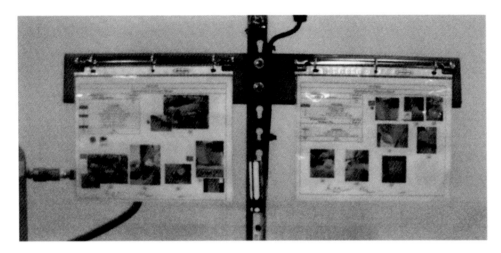

- Point of use
- Installed in workstation

Production control boards provide real-time status on daily progress of the work cell. They act as a scoreboard for the area, and workers are required to update the board as needed. Production control boards are intended to monitor output in small increments throughout the day. A variety of different types of boards exists, and the pace and volume of the work cell dictate the design. Here are the key attributes of a production control board:

- Required output
- Actual output
- Difference between required and actual
- Ongoing total
- Time-tracking interval
- Comments

The production control board shown next is an example of how to monitor output every hour. Our original Takt time for the 3R electric bike was 90 units a day in one shift. As you can see, the template is simple and easy to fill out.

Production Control Board					
Daily Rate	90		Date		
Time	Required	Actual	Difference	Total	Comments
8:00–9:00					
9:00–10:00					
10:00–11:00					
11:00–12:00					
12:00–1:00					
1:00–2:00					
2:00–3:00					
3:00–4:30					

Time: Monitoring interval

Required: Number units needed at the end of that hour

Actual: Number of units produced

Difference: Difference between required and actual

Total: Ongoing tracking of total output

Comments: Notes and issues during that hour

As you can see in the next image, the production control is filled. It is up to the worker in the last station to update the board every hour so information can be readily available. Resources can react at certain points when issues are placed in the "Comments" section. Keep track of this information as you can identify trends and make permanent improvements to the work cell.

Production Control Board					
Daily Rate	90			Date	
Time	Required	Actual	Difference	Total	Comments
8:00–9:00	9	9	0	9	
9:00–10:00	13	12	1	21	Tool Broke
10:00–11:00	10	10	0	31	
11:00–12:00	13	10	3	41	Waiting on Parts
12:00–1:00	6	6	0	47	
1:00–2:00	13	13	0	60	
2:00–3:00	10	11	+1	71	
3:00–4:30	16	16	0	87	

Required	90	Actual	87

You can quickly review this as a team every day in a stand-up meeting and discuss the board. As you can see, there were problems from 9:00 to 10:00 and 11:00 to 12:00. Also, the work cell workers were able to get back on track from 2:00 to 3:00. Place production control boards right in the work area, and they will become a valuable tool for tracking performance and getting to a root cause.

Conclusion

Cell manufacturing is an amazing improvement tool for any manufacturing process. You will find that not all processes are good candidates for work cells due to large, expensive, shared resources and if volume is just too low.

Never forget the value of good solid data up front and the importance of line balancing. Every work cell will be different, but the tools presented in this playbook will still apply.

Transitioning out of process-based production to cell manufacturing can be tough, and not everyone has the leadership skills or culture to make that change. I say this not to place doubt but to ensure you know what challenges you can face. However, it is making this leap to cell manufacturing that will catapult your company into becoming a much more competitive organization in your industry. Personally, I have implemented over 100 work cells, and the improvements in lead time, capacity, floor space, inventory, and cost have been amazing. Follow the tools in this playbook and you will find that your production lines are capable of so much more.

Good luck!

Definition of Terms

Available time: Time allocated in the day to work minus breaks, lunch, startup, and cleanup.

Communication lights: Used for many purposes; however, in a material replenishment system, this light is used to signal to material handlers that an order has been placed for more parts from their work area.

Daily walkthrough: Performed after the end-of-day cleanup, the walkthrough is conducted by a supervisor or worker to verify the cleanup is complete.

Defects: Mistakes made in the process requiring rework, material scrap, and lost products.

End-of-day cleanup procedure: A sustaining document that outlines the cleanup and reset requirements for the work area after each shift or day.

5S audit form: A scoring system used to rate the level of sustaining; used as a guideline for continuous improvement.

5S tracking sheet: A visual document posted in high-traffic areas that displays the scores from the 5S audit form.

5S and the visual workplace: Lean implementation concept of creating a highly organized work environment where everything has a place. Labels, designations, paint, and signage are examples used to create the visual workplace.

Floor space: Performance measurement of how much factory space is being used to conduct value-added work; often measured in profit per square foot or revenue per square foot.

In-process Kanban: A visual reordering system used to trigger the need to move or build WIP (work in progress) through a factory.

Inventory: Higher-than-needed inventory levels due to excessive purchasing of raw material, overproducing WIP, and unsold finished goods. Inventory ties up working capital, takes up floor space, and adds to longer lead times.

Kanban: Japanese word meaning "signal."

Kanban card: A visual reordering card placed near parts and material that contains information on how to order.

Motion: Movement of workers generally leaving their work areas to find items unavailable in their work areas.

Overprocessing: The act of overperforming work steps, such as redundant effort extra steps.

Overproduction: The act of producing more product than necessary, performing work in the wrong order, and creating unneeded inventory.

Production control board: Visual status board placed in the work cell to monitor output.

Productivity: One of the six Lean metrics that is a measurement of worker's efficiency in a process; often is a comparison of the time allocated to perform work to the actual time the worker took to perform it.

Quality: Internal measurement of rework, scrap, and defects in a production process.

Red Tagging: An organized approach to sorting in which Red Tags are placed on items to designate them as unneeded. Red Tag items are placed in a staging area for permanent removal from the company.

Right sizing: Concept of customizing the work area to identify the minimum amount of space needed to store items.

Scrub: Act of cleaning and painting the work area to create a showroom condition.

Set in order: Act of complete organization of the company by which all items are given home locations.

Shadow board: A visual mechanism for organizing tools. Shadow boards provide instant feedback on home locations and missing tools and opens up floor space by eliminating the need for toolboxes and shelves.

Single-piece flow: A concept of one piece of WIP allocated to a workstation/person. It is to ensure good flow and focus.

Sort: Act of discarding and removing all unnecessary items from the work area.

Spaghetti diagram: A study of the places, distance, and time workers spend walking around looking for items to perform work.

Standardize: Act of creating consistency in the 5S implementation through guidelines for the visual workplace.

Sustain: The act of maintaining the work area after a 5S implementation.

Takt time: Time interval that represents how often a completed product must be done.

Throughput time: Time associated with all value-added and non-value-added time in a process. It is the time it takes material to get through the first and last steps of the entire factory, raw material to finished goods.

Time and motion studies: The collection of work steps and time in a work area to capture the current state of assembly time and to be used to pinpoint improvements.

Transportation: The movement of raw, WIP, and finished goods throughout the company.

Travel distance: Measurement of the physical distance product and workers go and the time associated with it. A long travel distance equates to longer lead times in the process.

2-Bin system: A visual Kanban system in which parts or material are placed in two bins. Once a bin is empty, the empty bin acts as a Kanban.

Waiting: When work comes to a stop due to lack of necessary tools, people, material, information, and parts. Wait time is often called queue time.

Wasted potential: Poor use of people, including skill sets not being utilized, wrong job placement, and workers consumed in wasteful steps.

Index

About the Author

Chris Ortiz is the founder and president of Kaizen Assembly, a Lean manufacturing training and implementation firm in Bellingham, Washington. Chris has been featured on *CNN Headline News* on the show *Inside Business* with Fred Thompson. He is the author of seven books on Lean manufacturing (see the list that follows).

Chris Ortiz is a frequent presenter and keynote speaker at conferences around North America. He has also been interviewed on KGMI radio and the *American Innovator* and has written numerous articles on Lean manufacturing and business improvement for various regional and national publications.

Kaizen Assembly's clients include industry leaders in aerospace, composites, processing, automotive, rope manufacturing, restoration equipment, food-processing, and fish-processing industries.

Chris Ortiz is considered to be an expert in the field in Lean manufacturing implementation and has over 15 years' experience in his field of expertise. He is also the author of the following:

Kaizen Assembly: Designing, Constructing, and Managing a Lean Assembly Line (Taylor and Francis, 2006), now in its second printing

Lesson from a Lean Consultant: Avoiding Lean Implementation Failure on the Shop Floor (Prentice Hall, 2008)

Kaizen and Kaizen Event Implementation (Prentice Hall, 2009); translated into Portuguese

Lean Auto Body (Kaizen Assembly, 2009)

Visual Controls: Applying Visual Management to the Factory (Taylor and Francis/Productivity Press, December 15, 2010)

The Psychology of Lean Improvements: Why Organizations Must Overcome Resistance and Change Culture (CRC Press and Productivity Press, April 2012), winner of the Shingo Prize for Operational Excellence in Research, 2013

The 5S Playbook: A Step-by-Step Guideline for the Lean Practitioner (Taylor and Francis/Productivity Press, October, 2015), *The Lean Playbook Series*

The Kanban Playbook: A Step-by-Step Guideline for the Lean Practitioner (Taylor and Francis/Productivity Press, December 2015), *The Lean Playbook Series*

The Cell Manufacturing Playbook: A Step-by-Step Guideline for the Lean Practitioner (Taylor and Francis/Productivity Press, February, 2016)

The Quick Changeover Playbook: A Step-by-Step Guideline for the Lean Practitioner (Taylor and Francis/Productivity Press, March, 2016)

The TPM Playbook: A Step-by-Step Guideline for the Lean Practitioner (Taylor and Francis/Productivity Press, February, 2016)